LITTLE BOOK FOR HEART AND BLOOD VESSEL HEALTH

What is my risk for heart attack or another vascular event?

How do I achieve goal?

Philip H. Frost, M.D.

Little Book For Heart And Blood Vessel Health
What is my risk for heart attack or another vascular event?
How do I achieve goal?

Health & Fitness / General

ISBN: 145282116X

EAN-13: 9781452821160

WHY THIS LITTLE BOOK?

This Little Book is written to address an important problem. Atherosclerosis (hardening of the arteries) is the foundation for heart attack (myocardial infarction), stroke (cerebrovascular event) and diseases of the lower extremities and kidney circulation (together cardiovascular events). Heart attack is the most common cause of death in the United States and stroke is the leading cause of disability.

 Important to understand, is that these real problems can for the most part be prevented. The premise of this book is that a basic understanding of the lipid abnormalities present (the foundation for trouble), and the multiple treatment modalities available to correct such, will empower an individual to dramatically reduce the risk of suffering a heart attack or stroke (cardiovascular event).

This little book uses real examples to demonstrate what it takes to lead a likely cardiovascular event-free life. Success requires establishing a sensible lifestyle, attention to blood pressure, and the topic of this little book - use of medications to normalize evident lipid (cholesterol) disorders.

An additional comment: A passing lane overview is provided in the event that the reader wishes not to embrace the technical aspects of a chapter but is interested in the primary message. To accommodate formating, this is added as Appendix II. That said, this Little Book is a synopsis of a vast body of information and first choice is to read the chapter.

BASIC PHYSIOLOGY:

To begin, we need to briefly discuss two terms commonly used to describe lipid physiology, cholesterol and triglycerides. Cholesterol is a waxy insoluble substance and a required constituent of all animal cells. It is found in cell membranes and can be thought to serve a structural function analogous to a 2 X 6 inch timber in a wood framed building. Triglycerides, the chemical name for "fats," are storage molecules for energy. They contain three fatty acids. Fatty acids are "burned" for energy but, like cholesterol, serve additional functions. Cholesterol and triglycerides have one physical characteristic in common. Neither is soluble in water. Thus, unlike salts and proteins, cholesterol and triglycerides are not soluble, per se, in blood plasma. This leads to a potential transportation problem. This can be illustrated in the kitchen in the preparation of an oil and vinegar salad dressing. We begin with olive oil (triglycerides) and vinegar (water based, as is our blood plasma). After shaking the mixture but then pausing, we see that what appeared to be a solution is now in two phases. The oil has separated from the vinegar and is floating on top. Two conclusions: (1) triglycerides are not water soluble, and (2) triglycerides are lighter than water. This is true for cholesterol as well. Problem: how can these water insoluble substances be transported? Solution: construct lipoproteins.

Why Lipoproteins?

Architecture

Think of lipoproteins as spherical transporters. In organization, they have a surface and a core. Given that the storage form of fatty acids (triglycerides) and the storage form of cholesterol (cholesterol esters) are water insoluble, their locus for transportation is in the core of families of spherical lipoproteins. Lipoprotein families have thus a surface composed of constituents that are both water and fat soluble (amphipathic - like common soap or detergents) and a core containing the water insoluble triglycerides and cholesterol esters (Figure 1). For this discussion we will briefly consider six lipoprotein families (Figure 2). Each family is distinctive in size and additionally is distinguished by its surface proteins. The surface constituents are heavy relative to those in the core. Thus the largest lipoproteins are lighter or less dense than the smallest lipoproteins. Their names are for the most part reflective of the size or density of each lipoprotein family. These points are illustrated in the Figures 1 and 2 and Table 1. Table 1 provides component densities (weight relative to water), definition of commonly used acronym names, and details of size and composition for the lipoprotein families. The surface is 21 Angstroms thick. Surface proteins differ between and within lipoprotein families with details beyond the scope of this Little Book. This said, the primary apolipoprotein (protein) on chylomicron is B-48; on VLDL, IDL, LDL and Lp(a) is B-100; and on HDL, A-1.

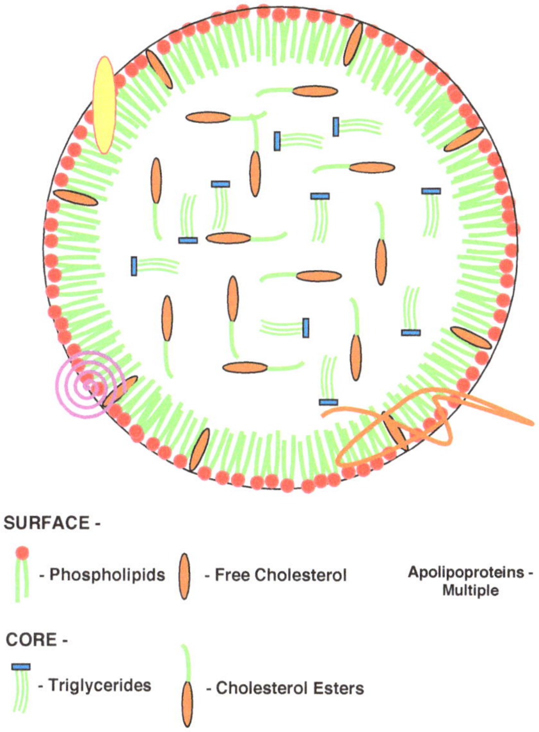

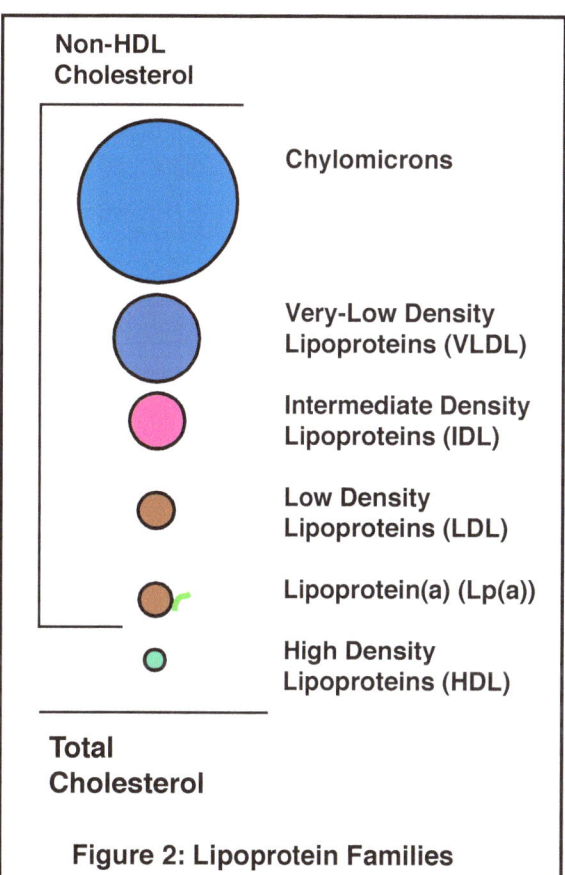

Figure 1: Lipoprotein and its Constituents

SURFACE -

- Phospholipids   - Free Cholesterol   Apolipoproteins - Multiple

CORE -

- Triglycerides   - Cholesterol Esters

Non-HDL Cholesterol

Chylomicrons

Very-Low Density Lipoproteins (VLDL)

Intermediate Density Lipoproteins (IDL)

Low Density Lipoproteins (LDL)

Lipoprotein(a) (Lp(a))

High Density Lipoproteins (HDL)

Total Cholesterol

Figure 2: Lipoprotein Families

Table 1. Size and Chemical Composition of Human Plasma Lipoproteins						
	Chylo-microns	**VLDL**	**IDL**	**LDL**	**HDL-2**	**HDL-3**
Diameter Å	800-5000	300-800	250-350	216	100	75
Density (gm/ml)	0.93	0.95-1.006	1.006-1.019	1.019-1.063	1.063-1.126	1.126-1.210
Percent Composition						
SURFACE						
Protein	2	8	19	22	40	55
Phospholipids (PL)	7	18	19	22	33	25
Free cholesterol (FC)	2	7	9	8	5	4
CORE						
Cholesterol-esters (CE)	3	12	29	42	17	13
Triglycerides (TG)	86	55	23	6	5	3
% Surface	11	33	47	52	78	85
% Core	89	67	52	48	22	16

VLDL - very low density lipoproteins
IDL   - intermediate density lipoproteins
LDL   - low density lipoproteins
HDL - high density lipoproteins
Å   - Angstroms
gm/ml - grams per milliliter
Reference #1

Density (gm/ml)

Protein - 1.33
PL   - 0.97
FC   - 1.06
CE   - 0.99
TG   - 0.915

Function

Lipoproteins serve distinct roles in transporting triglycerides and cholesterol. This is portrayed in Figure 3. Chylomicrons are assembled in the intestine and transport dietary fat and cholesterol into the plasma blood compartment. VLDL similarly

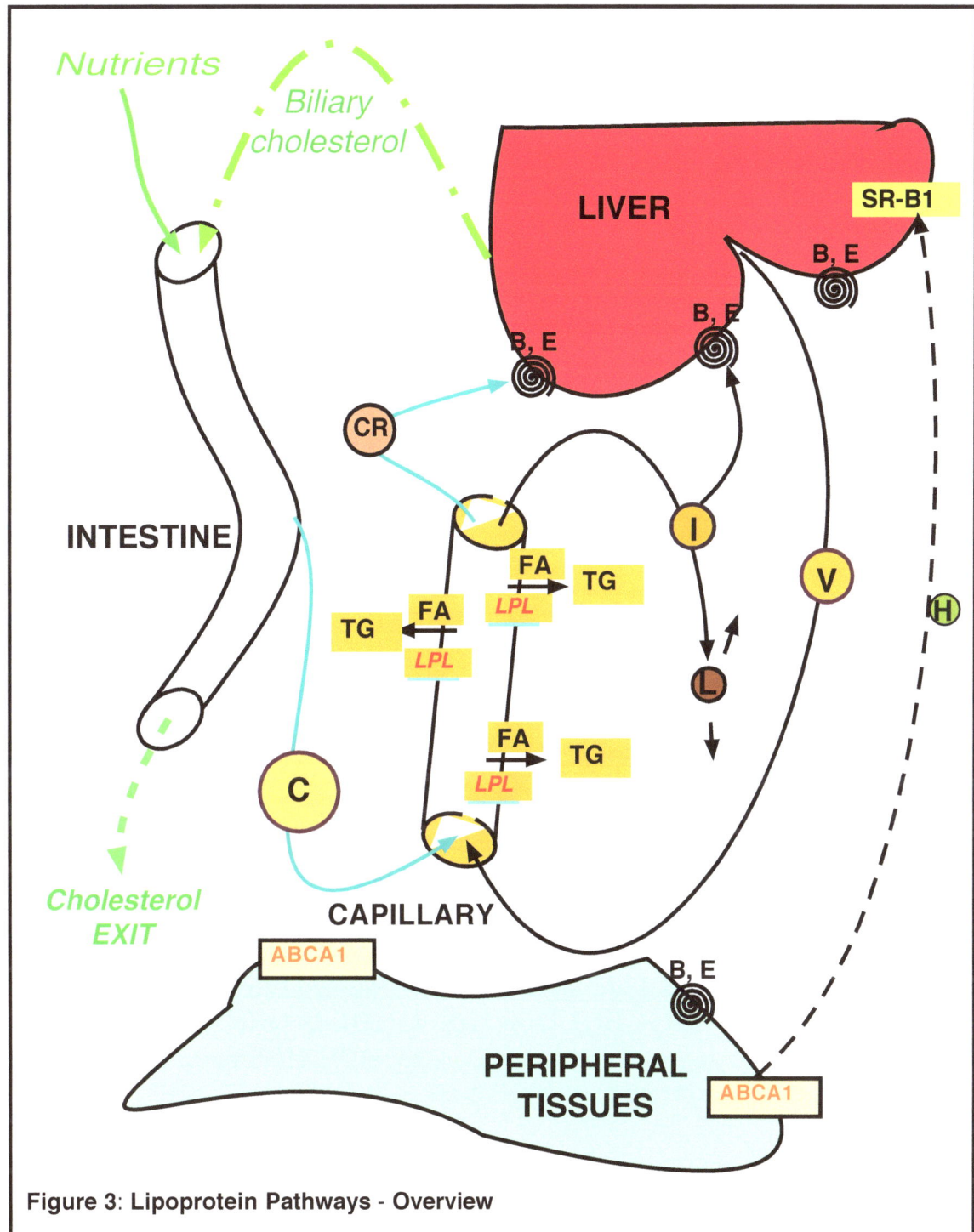

Figure 3: Lipoprotein Pathways - Overview

ABCA1 - receptor governing cholesterol efflux; B, E - LDL receptor recognizing apo B & E; C - chylomycron; CR - chylomicron remnants; FA - fatty acids; H - HDL; I - IDL; L - LDL; LPL - lipoprotein lipase; SR-B1 - receptor recognizing HDL; TG- triglycerides; V - VLDL transport triglycerides and cholesterol from the liver into the plasma compartment. IDL, LDL, and Lp(a) are derived from VLDL. HDL in contrast, transports cholesterol from non-liver cells including those in the artery to the liver, a process called reverse cholesterol transport. As noted above, VLDL and its offspring are characterized by a surface protein called apolipoprotein B-100. Chylomicrons are distinguished by apolipoprotein B-48, and HDL by apolipoprotein A-I (apo B-100, B-48, and A-I, respectively). Important is the fact that apo A-I bearing families are considered blood vessel friendly while the apo B carrying lipoproteins in high number can initiate and aggravate existing blood vessel disease.

This is too complicated, HELP!

 Often we simplify the story for clinical purposes: Routinely we estimate lipoprotein concentrations based on measured total cholesterol, total triglycerides and the cholesterol in HDL. Lp(a) lipoprotein is measured initially and, if elevated, in follow-up. The VAP test (Vertical Auto Profile) separates lipoproteins by density and provides a more detailed analysis. Blood is drawn after a 10 hour fast to avoid the contribution of chylomicron lipids. LDL-cholesterol as estimated (calculated) routinely includes that in true LDL, IDL, and Lp(a) [cover reference #2].   The cardiovascular risk associated with elevated cholesterol is commonly considered that associated with LDL-cholesterol. However there is also risk associated with VLDL-cholesterol. This combined risk is best described with non-HDL cholesterol (total minus HDL cholesterol). See Figure 2. Non-HDL-cholesterol includes the cholesterol in all apo-B containing lipoproteins. **Risk associated with elevated cholesterol is best summarized as non-HDL cholesterol.** However, risk assessment is not so simple. The cholesterol in HDL, apo A-I families, explains risk inversely. High HDL-cholesterol usually connotes lower risk whereas low HDL-cholesterol carries important independent risk. **Cardiovascular risk associated with lipoprotein cholesterol abnormalities is best assessed calculating the ratio of the non-HDL cholesterol to that in HDL (non-HDL-C/HDL-C)** or the ratio of apo B-100 to apo A-1 (when available). The ratio of total to HDL cholesterol gives the same information as ratio of non-HDL to HDL cholesterol (ratio of total to HDL cholesterol is one digit larger).

RISK ASSESSMENT (What is the likelihood That I Will Have a Vascular Event?

What are my chances of suffering a heart attack, stroke, or another atherosclerotic vascular event? Factors important to consider include family history, age, cholesterol disorders, blood pressure, cigarette smoking status, presence or absence of diabetes, male sex, and other factors, such as evident coronary artery plaque and measures of oxidative stress (inflammation) [examples are high-sensitive C reactive protein (hs CRP) and lipoprotein-associated phospholipase A_2 (Lp-PLA$_2$)].

Risk, as commonly described, is short-term risk: the likelihood of an event within a 10-year time frame. We need to consider both short-term and long-term risk.

Risk is best illustrated with figures. The differences in short-term and lifetime risk of coronary heart disease at selected ages is illustrated in Figure 4. For simplification, data presented are limited to ages 40 and 60. Note the distinct difference in short-term and lifetime risk, both cholesterol driven. We need to think about short-term and lifetime risk in considering initiation of risk factor reduction. Figure 5 describes risk of coronary heart disease (CHD) death within 25-years across a broad range of total cholesterol (<160 to >280). CHD death approximately doubles in steps of 40 mg per deciliter. Figure 6 assesses cardiovascular disease (CVD) risk with simple risk factors: total cholesterol, blood pressure, smoking and diabetes status. These data from the Framingham Heart Study are eye popping in regard to risk associated with traditional risk factors. Figure 7 demonstrates that conventional risk factors define not only the chance of CVD death but also non-CVD death. Simple measurement of blood pressure, total cholesterol, smoking and body weight (described as body mass index [derived from weight and height]) predict age related death from all causes.

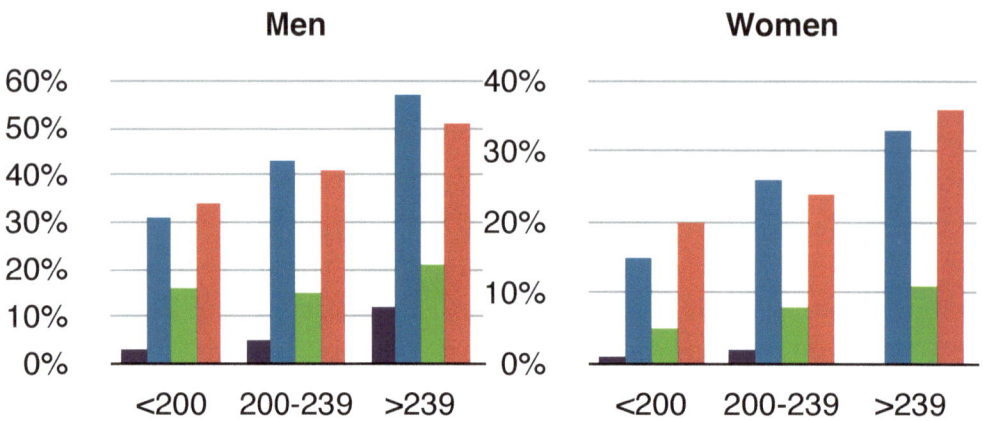

Figure 4. Ten-Year and Lifetime Risk of Coronary Heart Disease by Cholesterol Levels at Selected Ages. The data demonstrates the usefulness of considering absolute life-time risk compared with shorter-term risks. For younger participants, the short-term risk of CHD is exceedingly low even with elevated cholesterol levels, whereas the lifetime risk is high, especially for those with elevated cholesterol level. Data from the Framingham Heart Study. Reference #2.

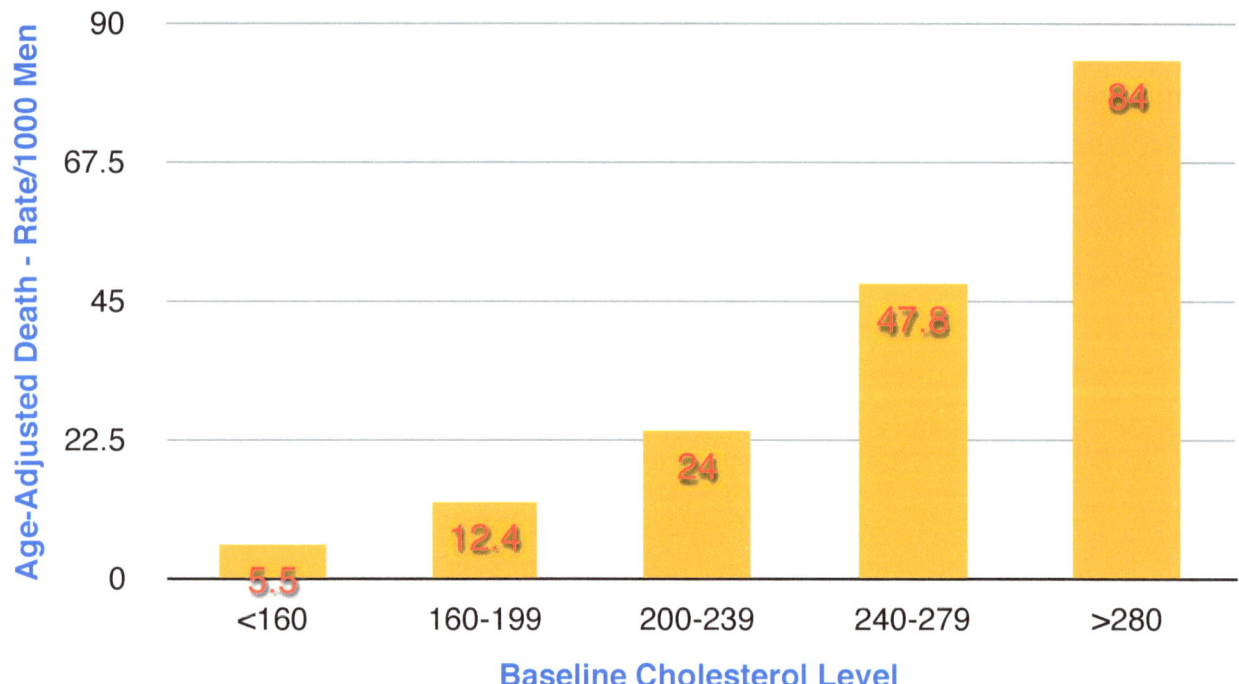

Figure 5. Risk of CHD Death in Young Men: Chicago Heart Association - 11,017 Men Aged 18 to 39 at Baseline - 25-Year Follow-up. Reference #3.

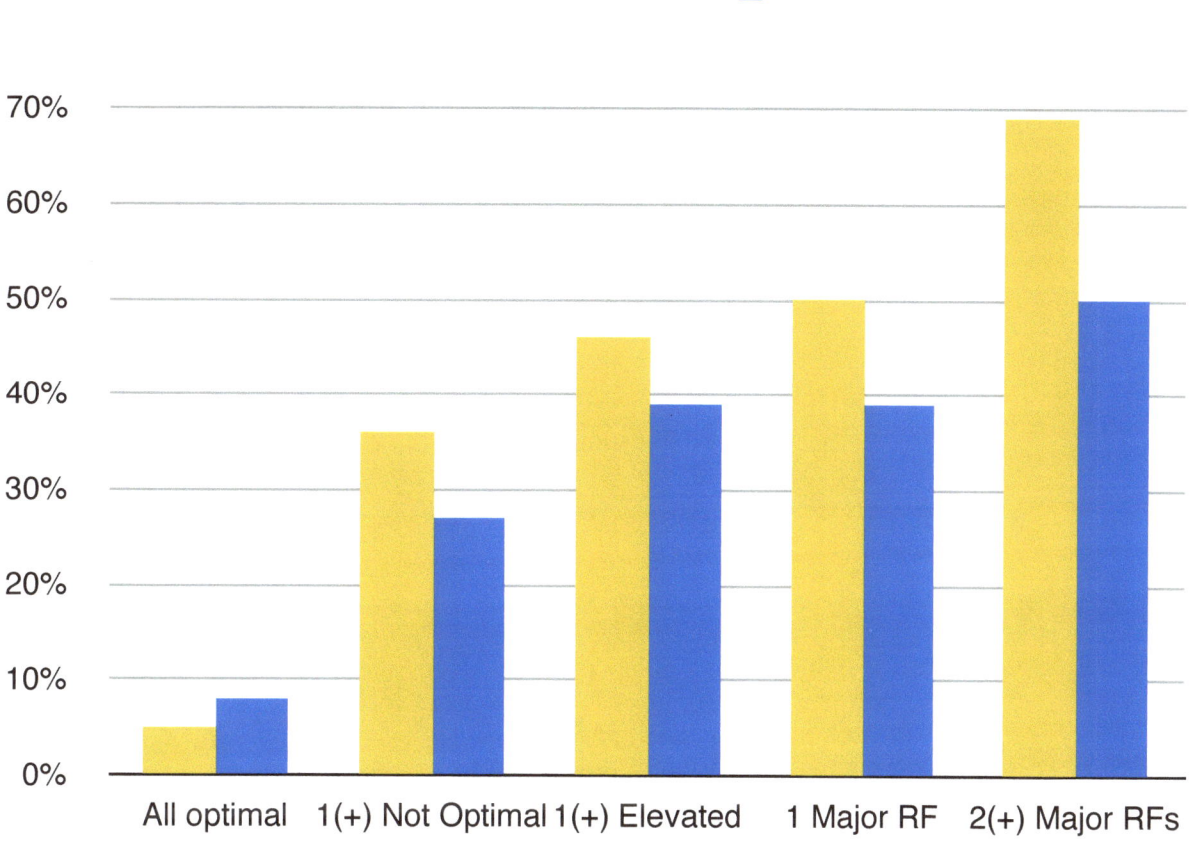

Figure 6. Lifetime risk for CVD (to age 95) for men and women with *all optimal risk factors* [total cholesterol < 180, blood pressure <120/<80 mm Hg, nonsmoker, and non diabetic]; *1 or more non optimal RFs* [TC 180-199], systolic blood pressure 120-139, diastolic blood pressure 80-89], nonsmoker, nondiabetic; *1 or more elevated RFs* [total cholesterol 200-239, systolic blood pressure 140-159, diastolic blood pressure 90-99, nonsmoker and nondiabetic; *1 major RF elevated*, *2 or more major risk RFs elevated* [major risk factors are defined as total cholesterol >239, systolic blood pressure >159, diastolic blood pressure >99, smoker, and diabetic. Data are from the Framingham Heart Study. Reference #4.

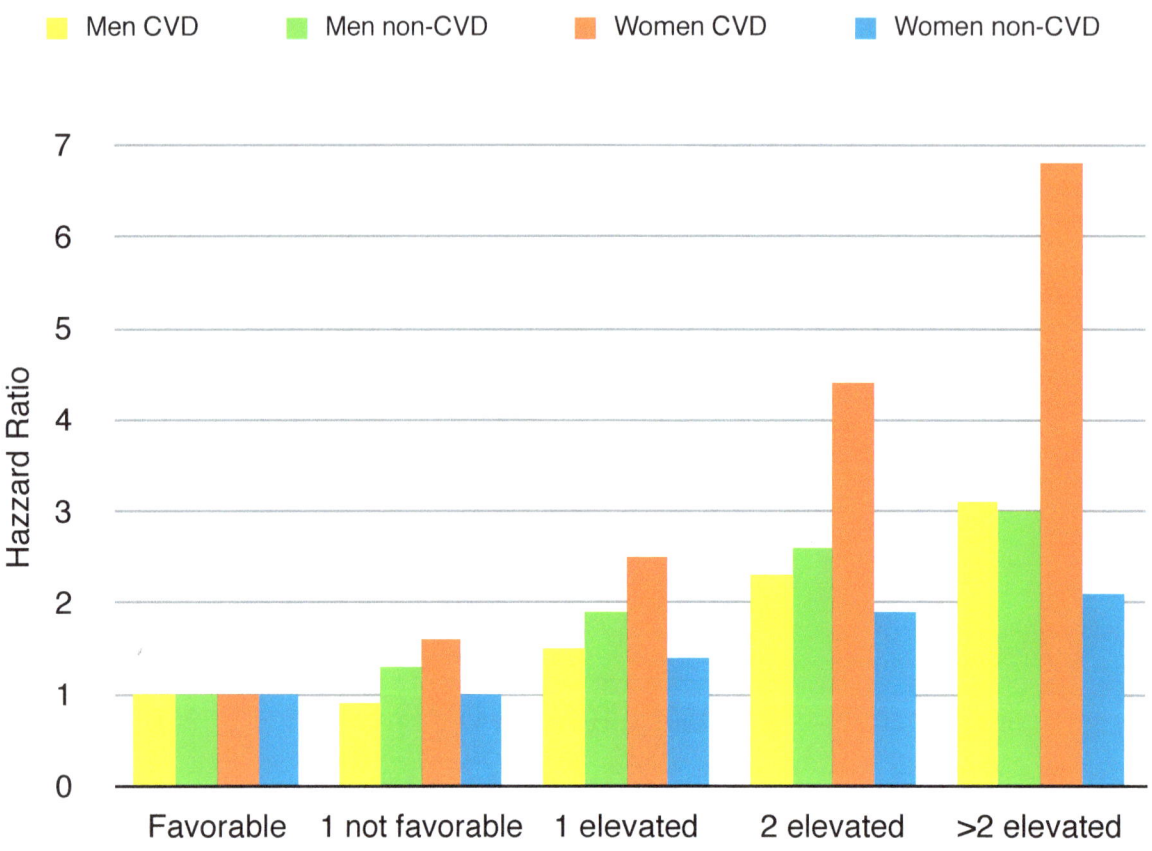

Figure 7. Age adjusted hazard ratios for cardiovascular disease and non-cardiovascular disease death by aggregate risk factor (RF) burden in middle age (aged 40-59 - average follow-up 32 years). *Favorable RF profile* [systolic blood pressure (SPB) <120 and diastolic blood pressure (DBP) <80 mmHg, and total cholesterol <200 mg/dL, and non-smoking, and body mass index <25 kg/m²]; ***1(+) Not Favorable*** [no elevated risk factor, but 1 or more risk factors not at favorable levels]; ***1 RF elevated***; ***2 RFs elevated***; ***>2 RFs elevated*** [elevated risk factors include SPB >140 or DPB>90 mm Hg or receiving anti-hypertensive therapy, total cholesterol >239 mg per deciliter, current cigarette smoking, or body mass index greater than 30 kg/m²]. Data from The Chicago Heart Association Detection Project in Industry. Reference #5

These long-term studies utilized available total cholesterol for determination of lipid associated risk. Today when we consider an individual's risk, we have the advantage of modern lipoprotein measurements. At the same time, individual risk is not always definable with current clinical measurements. Family history when available is often helpful in risk assessment. Updated risk assessment tools are clearly needed. But, critical to the message of this Little Book, is the fact that modifying currently identifiable risk factors has clear benefit.

WHY DO I HAVE ABNORMAL LIPOPROTEIN LEVELS?

Lipoprotein levels are governed by both primary and secondary causes.

Primary cause(s)

Primary cause(s) (a function of genetic makeup) may be expressed in a dominant fashion, clear transmission from generation to generation, and/or in a recessive manner, where each parent contributes a gene that may not be evident. Eye color is an easily visualized example of dominant and recessive transmission. A brown-eyed father and blue-eyed mother have two children: one with brown and the other with blue eyes. The brown-eyed child inherited the father's dominant brown-eyed gene and mother's blue-eyed gene. The blue-eyed child inherited blue-eyed genes from both parents (father has hidden blue-eyed gene). To be blue-eyed you need 2 blue-eyed genes. To be brown-eyed, one brown-eyed gene is sufficient. If both parents are heterozygous for eye color (one brown- and one blue-eyed gene), there is a one of four chance they could have a blue-eyed child. The blue-eyed gene transmission is recessive - not visible in either parent.

Secondary causes

Secondary causes include lifestyle factors and the presence of ancillary medical confounders. Lifestyle factors include food choice, physical activity, obesity, stress, and alcohol and cigarette use. Medical confounders include the presence of diabetes and it's control, hypothyroidism, kidney disease, and biliary obstruction of the liver. Medications prescribed for other purposes may also modulate lipid levels. The bottom line is that lifestyle factors need to be addressed clearly and secondary medical disorders need to be diagnosed and corrected to the degree possible. Treatment of a hypothyroid patient with thyroid replacement will improve (sometimes correct) his lipid disorder. A complete discussion of the genetic and secondary causes of lipoprotein disorders, their diagnosis and treatment, is beyond the scope of this Little Book.

HOW DO PATIENTS PRESENT?

Patients present with a wide range of lipoprotein abnormalities and family and personal histories. If one walks through a neighborhood, one sees ordinary people with ordinary problems. These are the people who as a group have most events. If one works in a referral center, one sees people who often have distinct genetic abnormalities placing them at high risk. These individuals may comprise 1-3% of those in the neighborhood. All need to consider their risk, and move to goal to avoid or delay the development of CVD, the primary killer in the US.

Three tables illustrate the fact that not all patients look alike. Table 2 summarizes the lipid measurements of 490 patients who when first seen were not taking lipid altering medications. Note the ranges at presentation. Table 3 expands these summary data. Note that within each step-range of cholesterol, there is marked heterogeneity in other lipids. Table 4 illustrates the group and three members of the group who presented with a similar total cholesterol - that is within the range of 190-210. Note the ratios of non-HDL to HDL cholesterol. With similar total cholesterols, there are multiple lipid disorders (phenotypes) to resolve.

Table 2. Not All Patients Look Alike - Patients Who When Referred for Evaluation and Treatment Were Not taking Lipid Altering Agents (n=490)				
	TC	**TG**	**HDL**	**non-HDL**
Ave	249	297	51	198
Min	90	32	5	63
Max	1157	4820	145	1072
SD	95	439	20	93
TC = total cholesterol, TG = total triglycerides, HDL = HDL cholesterol, non-HDL = non-HDL cholesterol (TC - HDL), Ave = average, Min = minimum, Max = maximum, SD = standard deviation				

Table 3. Referred Patients - Information Displayed by Step Ranges of Total Cholesterol, 150-199, 200-249, etc.

	TC	TG	HDL	non-HDL
TC 150-199				
Ave	176	195	44	132
Min	152	38	20	88
Max	199	1110	103	174
TC 200-249				
Ave	225	215	51	174
Min	200	50	17	112
Max	249	979	99	232
TC 250-299				
Ave	271	259	58	214
Min	250	32	21	143
Max	299	1082	123	273
TC 300-349				
Ave	320	398	57	265
Min	300	42	5	183
Max	347	2709	145	326
TC ≥350				
Ave	453	989	56	398
Min	354	82	14	250
Max	1157	5028	128	1072

For Abbreviations see Table 2 - Not All Patients Look Alike

Table 4. Total Cholesterol (TC) Does Not Tell the Whole Story - Group and Three Illustrative Patients Who Presented with TC Ranging From 190-210

	TC	TG	HDL	LDL	non-HDL	non-HDL/ HDL
Mean **n = 57**	200	204	49.5	114	151	3.6
Patient A	191	110	103	66	88	0.9
Patient B	204	230	25	133	179	7.2
Patient C	195	498	35		160	4.6

TC - total cholesterol, TG - total triglycerides, HDL - high-density lipoprotein cholesterol, LDL - low-density lipoprotein cholesterol, non-HDL - cholesterol not in HDL, non-HDL/HDL - ratio non-HDL/HDL cholesterol. LDL cholesterol can not be properly estimated when TG is >400 mg/dL.

WHAT IS GOAL?

Multiple organizations have issued guidelines for the treatment of individuals with lipid disorders. Table 5 summarizes the recommendations of the National Cholesterol Education Program (NCEP) establishing goals for LDL cholesterol and non-HDL cholesterol by risk category. NCEP emphasizes establishing treatment and goals based on the 10-year likelihood of developing a cardiovascular event. We are interested in strategies to reduce both lifetime and short-term risk of suffering a cardiovascular event.

Is there additional information that we should consider in a discussion of setting goals and considering treatment strategies? Table 6 summarizes data from an experiment in nature. Individuals with specific DNA-sequence variations (in the gene PCSK9) have a lifelong lower plasma LDL than their neighbors. These sequence variations differ in white and black subjects with consequent LDL differences in blacks more striking. In individuals screened at ages 45 to 64 years and followed for 15 years there was a real difference in the coronary heart disease event rate in the groups with lifelong lower LDL cholesterol levels. Other CVD risk factors were not different. Here we see the benefit of long-term lower LDL levels.

Table 7 displays the author's suggested goals for people with lipid disorders. I will often say - well below these levels. Not considered in Table 7 is elevated lipoprotein (a) [Lp(a)]. Some individuals have aggressive blood vessel disease related to elevated Lp(a) and this need be addressed in designing a treatment regimen. The knowledge database that is the foundation for defining goals is continuously changing. Keep informed!

Table 5. National Cholesterol Education Program's Goals for the Treatment of Individuals with elevations of LDL or Non-HDL Cholesterol (Non-HDL cholesterol is a secondary target for those with TG >200)

Risk Category	LDL-C	Non-HDL-C
CHD, or PVD, or diabetes, or CHD risk equivalent (10-year-risk > 20%)	<100	<130
Multiple risk factors & 10-year-risk < 20%	<130	<160
0-1 risk factors	<160	<190
Reasonable Goals (for high risk individuals)	<70	<100

Risk categories are based on presence of (1) coronary heart disease (CHD), (2) peripheral vascular disease [including cerebral vascular disease] (PVD), (3) diabetes mellitus, (4) Framingham Risk Score (calculated), (5) established "risk factors": a. cigarette smoking; b. hypertension (BP ≥ 140/90 or on antihypertensive medication); c. low HDL cholesterol (<40 mg/dL); d. family history of premature CHD [parent or sibling - male < 55 years, - female < 65 years]; e. age - male ≥ 45 years, female ≥ 55 years. References # 6 and #7.

Table 6. Sequence Variations in PCSK9, Low LDL and Protection Against Coronary Heart Disease								
	% with Mutation	No Mutation			Mutation			
		TC	LDL-C	% CHD	TC	LDL-C	% CHD	
White Subjects	3.2%	214	137	11%	194	116	7%	
Black Subjects	2.6%	215	134	10%	173	100	1%	

Participants in the "Atherosclerosis Risk in Community Study" were screened at ages 45-64 years and followed for 15 years. Amongst white subjects, 3.2 % (301 of 9524) had a PCSK9 mutation, a 15% lower LDL cholesterol (116 vs 137) and 47% lower CHD incidence (7 vs 11 %). Amongst black subjects, 2.6% (85 of 3363) had a PCSK9 mutation, a 28% lower LDL cholesterol and a 88 % lower CHD incidence. Reference #8.

Table 7. Lipid Goals - Author's Perspective					
	Non-HDL	LDL	Non-HDL/ HDL ratio	HDL	TG
Goal 1	<130	<100	<3.0	>40	<150
Goal 2	<100	<70	<2.0	>50	<120

Goal 1 is suggested for individuals who do not have recognized blood vessel disease. Goal 2 is suggested for those at high risk and those with established blood vessel disease. The non-HDL and LDL cholesterol goals are recognized goals. The ratio of non-HDL to HDL and HDL cholesterol and TG goals defined are likely to become standard. NOTE: The ratio sometimes described is TC/HDL this differs from non-HDL/HDL by one unit.

HOW DO WE ACHIEVE GOAL?

To achieve goal, we need to be attentive to hygienic measures and consider a proper medication regimen. Hygienic measures extend beyond a thoughtful diet, weight management, and sustainable exercise activities, to include attention to alcohol intake, ideal diabetes control, and avoidance of medications that adversely affect lipids. Alcohol intake leads to obligatory increase in VLDL secretion by the liver and thus consequent increased blood triglycerides and often downstream lipoprotein elevations. Diabetes not ideally controlled has similar consequence. Weight management is rewarded with decreased VLDL synthesis and thus improvement in most lipoprotein abnormalities. Oral estrogens increase VLDL secretion but, if needed, can be prescribed transdermally (by patch) with lesser lipoprotein perturbation. Cigarette smoking lowers HDL cholesterol in addition to multiple other adverse effects. Discussing the fine points of hygienic modification is beyond the scope of this Little Book.

To achieve goal, medications are almost always required.

What medications are available for the treatment of lipoprotein disorders?

Multiple classes of medications are currently employed to achieve lipoprotein perturbation. They are listed below in the order they were discovered to perturb lipids historically (Niacin was noted to have lipid effects in 1955.). Phytosterols (plant sterols) were recognized to reduce blood cholesterol in the 1980's but, for simplicity purposes, are here included in the category cholesterol absorption inhibitors.

1. Niacin (nicotinic acid)
2. Fibrates
3. Bile acid sequestrants
4. Statins
5. Omega-3 (n-3) fatty acids
6. Cholesterol absorption inhibitors

The essential information regarding these drug classes is presented in Tables 8 and 9. Table 8 lists drug by class with primary mode of action. Table 9 summarizes likely lipoprotein response and potential side effects for agents listed. Response to each maneuver of course is individual.

Table 8. Lipid Active Agents

Class	Drug	Primary Mode of Action
Niacin (vitamin B3)	Niacin – Rapid release capsules or tablets Niaspan	Multiple
Fibrates	Fenofibrate Gemfibrozil	Improve VLDL clearance; decrease VLDL production
Bile Acid Sequestrants	Cholestyramine Colestipol Colesevelam	Capture bile acids in the intestine, stimulating production of replacement bile acids from liver cholesterol
Statins	Lovastatin Pravastatin Simvastatin Fluvastatin Atorvastatin Rosuvastatin	Inhibit cholesterol synthesis primarily in the liver, up-regulating B,E receptors and enhanced clearance of LDL and IDL by the liver
Omega-3 Fatty Acids	EPA + DHA	Decrease VLDL synthesis in the liver
Intestinal Cholesterol Absorption Inhibitors	Ezetimibe Plant Stanols	Decrease intestinal cholesterol absorption, leading to reduction in transport of cholesterol to the liver, and enhanced liver clearance of LDL and IDL

Table 9. Likely Response to Lipid Active Agents

Drug	Lipoprotein Effects				Side Effects
	⇓ VLDL	⇓ LDL	⇓ Lp(a)	⇑ HDL	
Niacin – Rapid Release	⇓⇓⇓	⇓⇓	⇓⇓	⇑⇑⇑	Flush, elevate uric acid; may increase liver enzymes, blood glucose; occasional dry skin; rare acanthosis nigricans, macular edema
Niaspan	⇓⇓	⇓	⇓	⇑⇑	As above for niacin
Fenofibrate Gemfibrozil	⇓⇓	0	0	⇑	May increase liver enzymes; may increase gall stones; muscle syndrome, rare
Cholestyramine Colestipol	⇑	⇓⇓	0	0	Possible GI distress, constipation
Colesevelam	⇑	⇓⇓	0	0	Possible GI distress
Pravastatin Fluvastatin	⇓	⇓⇓	0	⇑	Muscle syndrome, rare; may increase liver enzymes
Lovastatin Simvastatin	⇓	⇓⇓⇓	0	⇑	As above for other statins
Atorvastatin Rosuvastatin	⇓	⇓⇓⇓⇓	0	⇑	As above for other statins
EPA + DHA	⇓⇓	0	0	0-⇑	None
Ezetimibe	0	⇓⇓	0	0	Possible GI distress; muscle syndrome, rare
Plant Stanols	0	⇓	0	0	None
Members of drug classes are here distinguished (separated) secondary to expected efficacy or potential side effects (see drug specific comments below]. IMPORTANT: These are crude estimates of likely change. Individual responses are quite variable and easily discerned by repeated blood testing.					

Drug specific comments

The comments below are a reflection of the author's 40-year experience as a clinical investigator and prescribing physician treating patients with lipid disorders. They are not inclusive nor comprehensive.

Niacin (nicotinic acid) at low dose is vitamin B3 and at high dose is a potent lipid modifying agent. Niacin is prescribed in one of two preparations: immediate release [IR] or sustained release (Niaspan). The longstanding experience with niacin (first noted to lower blood cholesterol in 1955) is with IR preparations. Immediate release niacin is available in tablet or encapsulated forms (capsules) and is available over-the-counter. IR niacin is used in doses ranging from 1.5 to 6 g daily. Niaspan is available by prescription, the usual dose is 0.5 to 2 g taken at bedtime. The potential for liver problems (hepatotoxicity) is in part related to formulation. The release characteristics of the encapsulated product is predictable where at times a tablet preparation has slower release characteristics. As a result, patients who have developed hepatotoxicity with tablets may do well with the encapsulated product. The encapsulated product is thus preferred. Niaspan in contrast can be used safely at 0.5 to 2 g total dose only (usual dosage 1-2 g). Niacin as indicated above (Table 9) is the all-purpose lipid modifying agent leading to (1) decrements in the levels of all atherogenic lipoprotein species including Lp(a), (2) increased HDL concentrations, and (3) modification of LDL subtractions (shifting diameters from small dense to more buoyant LDL - thought to be less atherogenic). Niacin administration is accompanied by a nearly inevitable flush (see below) and an increase in blood uric acid (~1 mg/dL). Patients need to be monitored with liver function studies and fasting glucose. Elevated liver function studies may reflect hepatotoxicity and require dosage adjustment. In patients with diabetes or a propensity for diabetes, glucose may increase (usually transitory) but, rarely, type 2 diabetes mellitus can be precipitated. On occasion dry skin occurs, rarely pigment darkening and roughening of the skin in axilla (acanthosis nigricans), and very rarely vision change consequent to macular edema. All side effects and toxicities reverse with drug discontinuation. Important is the uncomfortable but harmless side effect of flush. This is a result of dilatation of the small blood vessels in the skin usually most prominent in the face and upper body. Important to know is that the flush usually abates when doses are increased to an individual's threshold level, often above 2 to 3 g per day. Taking the drug with meals helps. Pretreatment with aspirin or NSAIDs modifies the flush. My recommendations for initiating niacin and dosage advancement are expanded in appendix 1. In terms of cost, the immediate release encapsulated product is inexpensive. Niaspan is more costly.

The fibrates (gemfibrozil and fenofibrate) are second-tier agents employed for reductions of VLDL levels. They are only modestly effective in raising HDL cholesterol. Clinical trials to date show mixed results with decrement in heart attack but no benefits in terms of overall clinical outcomes. Fenofibrate but not gemfibrozil can be used in combination with a statin at low-dose. They are available in generic and branded preparations.

The bile acid sequestrants (cholestyramine, colestipol, colesevelam) are second tier agents employed for the reduction of LDL cholesterol. They have no HDL effect and may increase VLDL (triglycerides). LDL reduction is a consequence of sequestering bile acids in the intestine (normally reabsorbed [recycled]) requiring replacement from existing hepatic cholesterol stores. Bile acid sequestrants are not absorbed and potential side effects are limited to the gastrointestinal tract (constipation most common). Attention is required to dose timing for these agents may sequester not only bile acids but medications such as thyroxin or digitalis. Colesevelam is the best tolerated of these agents (but quite expensive).

The statins (lovastatin, pravastatin, simvastatin, fluvastatin, atorvastatin, rosuvastatin) are effective agents for lowering LDL, and to a lesser degree, VLDL cholesterol. They raise HDL modestly. These agents have been studied extensively in clinical trials with demonstrated reductions in heart attack, stroke and total mortality. Their safety profile is well-established with the primary side effect a rare muscle syndrome which is reversible with drug discontinuation. Lovastatin, the oldest statin, was approved and marketed in 1987. Lovastatin, pravastatin, simvastatin, and fluvastatin are now generic agents. Simvastatin is probably the best studied. Of concern to this author is that market share appears to be dominated by branded agents rather than those now generic and so well studied. The newer agents are more effective (in regards to LDL lowering). The generic agents are the least expensive.

EPA + DHA (eicosapentaenoic acid + docosahexaenoic acid) are long-chain omega-3 (n-3) fatty acids important in many metabolic pathways. When taken in doses of 3 g or more they have clear lipoprotein effects - decreasing VLDL (triglycerides) and modestly increasing HDL. They perturb lipids similar in degree to the fibrate class of drugs. EPA + DHA have multiple other potential benefits. These include potential to decrease incidence of arrhythmia, inhibit platelet aggregation (clotting), and modulate unwanted inflammatory responses. The primary dietary source of EPA and DHA is oily fish with other food sources less important. Fish consume micro algae - the source of these important fatty acids. EPA and DHA are available in capsule form (distillation purified) over-the-counter. Organic mercury compounds and PCBs (polychlorobiphenyls) are potential contaminants of marine fish oils. The product label should state "mercury free," distilled, or USP approved. It is important to realize that products differ in content of EPA + DHA. A prescription brand is available as Lovaza. The cost for 3+ grams daily varies wildly.

Ezetimibe is a direct inhibitor of cholesterol absorption from the intestine. As such, it inhibits both the absorption of cholesterol consumed in the diet and cholesterol which enters the intestine from the bile. Consequently the flux of cholesterol absorbed and packaged in chylomicrons and destined for the liver is reduced. In addition, the low blood concentration of plant sterols, also atherogenic, are reduced. In people who are moderate or high cholesterol absorbers, ezetimibe leads to a modest decrease in VLDL and a more significant decrease in LDL. Ezetimibe is prescribed alone and in combination with a statin. It is marketed both alone and in combination with simvastatin. Clinical trials with ezetimibe have not yet been completed and clinical benefit is not yet established.

Plant stanols when ingested in significant amounts lead to a reduction in cholesterol absorption by a mechanism distinct from ezetimibe. Cholesterol to be absorbed requires packaging. This process is inhibited partially by plant stanols. Plant stanols are available in concentrated form and in margarine preparations. Stanol use leads to a modest reduction in LDL cholesterol. Dietary fiber also reduces LDL cholesterol. The likely mechanism is that similar to the bile acid sequestrants (above). The long-term benefits of plant stanol ingestion are not established.

APPROACH TO DRUG PRESCRIPTION

Drug prescription is tailored to an individual's problem. Drugs may be prescribed as single agents or in combination. Many (maybe most) people will need a combination regimen to achieve lipid goals. Possible interactions between lipid drugs and other agents taken by the patient needs to be monitored. A detailed description of possible drug interactions is beyond the scope of this Little Book and need to be discussed with the prescribing physician.

When I use the word drugs, I am referring to lipid altering agents that are produced by both the pharmaceutical industry and mother nature. Niacin and the omega-3 fatty acids are found in nature where most statins are produced synthetically (lovastatin is isolated from a strain of fungus). Whatever its source, when prescribing a medication we consider not only lipoprotein change, but clinical outcome - demonstrated reduction in heart attack and stroke and, when data is available, the gold standard for clinical outcome - overall death rate (total mortality). For initial prescription, it seems logical to consider agents known to reduce cardiovascular events and improve total mortality. Agents known to improve overall mortality include selected older statins and niacin. Studies of the omega-3 fatty acids have been less formal but multiple benefits have been demonstrated. Clinical studies with ezetimibe are pending. Atorvastatin and rosuvastatin have demonstrated CVD benefits, but to date no data on total mortality is available. Studies of fibrates have not demonstrated reduction in total mortality.

WHAT CAN WE LEARN FROM PATIENT STORIES?

Large-scale clinical trials provide essential data in regards to efficacy and safety of the tested agent. Most clinical trials are funded by pharmaceutical companies and patients are selected who are likely to benefit from the agent being tested. Additionally, these trials, in almost all circumstances, test a single maneuver (usually a drug) for clarity. However, in real life, patients often present with problems not included in clinical trial populations and who need combined drug regimens.

Why review individual stories? It is likely that patient stories can help us fill knowledge gaps regarding maneuvers not studied in clinical trials. The stories outlined below have been chosen from the author's patient group to illustrate the power of hygienic measures in special circumstances and a variety of single and combination drug regimens. The reader of this Little Book is likely to have a lipid disorder and may find a like example useful in their quest to achieve lipid goal and reduce their likelihood of a clinical problem.

The patient stories are presented in a standard format with the final paragraph in italic font - lessons learned.

I: Man with known coronary artery disease and marked elevation of triglyceride-rich lipoproteins managed with attention to hygienic means.

History: This man was referred in September of 1985 at age 62. He had a 14-year history of hypertension and a previous coronary artery angioplasty procedure. He was referred for evaluation and management of known hyperlipidemia. He was a non-smoker and non-drinker. There was no clear family history of early coronary artery disease. He had gained 40 pounds since age 20 with weight stable in recent years. On physical examination he was moderately overweight but had no corneal arcus or xanthoma. Initial blood studies showed a total cholesterol 461, triglycerides 2360, and HDL cholesterol of 22. LDL cholesterol could not be estimated with a triglyceride greater than 400. His non-HDL cholesterol was 439. Blood glucose was elevated at 148 mg/dL (normal 70-100), hemoglobin A1c was elevated at 7.8 (test reflects average glucose level over preceding weeks [laboratory normal 4.8-6.7%]). See the table for response to treatment. Initial visit is highlighted in yellow.

Date	TC	TG	HDL	LDL	non-HDL	non-HDL/HDL	FBS	Hb-A1c	Wt	Med
09-85	461	2360	22		439	20.0	148	7.8	179	
10-85	205	151					117	5.4	168	
11-85	171	142	40	103	131	3.3	110	5.8	163	
01-86	164	122					98		157	
09-91	212	135	40	145	172	4.3	109	5.2	166	
05-92	159	116	47	89	112	2.4	106		168	L20

TC - total cholesterol, TG - total triglycerides, HDL - high-density lipoprotein cholesterol, LDL - low-density lipoprotein cholesterol, non-HDL - cholesterol not in HDL, non-HDL/HDL - ratio non-HDL/HDL cholesterol, FBS - fasting blood sugar, HbA1c - hemoglobin A1c, Wt - weight, Med - lipid medications, L20 - lovastatin 20 mg.

Follow-up: At his initial visit September 1985 he was advised to move to a low-fat, low-calorie diet and initiate weight loss. When his initial blood study results became available it was clear that he had not only marked hyperlipidemia but type 2 diabetes. Arrangements were made for him to procure a blood glucose monitor for twice daily monitoring. With the blood glucose data prompting food choice, he succeeded in losing 11 pounds when he returned in October. Blood glucose response was clear and lipid response was dramatic. Review of the data in the table clearly documents the lipid and blood glucose responses secondary to weight reduction (and exercise). In 1992, lovastatin was added in low dose with clear additional response.

In the fall of 1986 he underwent bilateral carotid artery (neck artery) surgery. In September of 1986 he underwent coronary artery bypass graft surgery. In June 1993 he underwent AV node ablation and pacemaker placement for an arrhythmia. This man has subsequently been followed remotely (he lives at a real distance) and he continues to do well with procedures limited to PTCA and stent placement 2001. Otherwise he has had no interval need for vascular surgery nor additional major medical problems. When contacted September 2008, he felt well, working out at a gymnasium 6 days a week. He is about to celebrate his 85th birthday.

Lipoprotein levels are governed by both production and removal rates. VLDL production is governed by both genetic and environmental factors. VLDL removal is likely genetically limited in patients who develop marked hypertriglyceridemia. Patients with type 2 diabetes secondarily have an increase in VLDL production and control of diabetes is crucial in managing the lipid disorder. In this case, the discovery and control of his diabetes was followed by a marked lipid benefit. Type 2 diabetes can often be managed with real weight reduction (and exercise). When weight control alone is not sufficient, medications are needed. In regard to lipid control, benefit is seen with weight reduction and exercise but medications are often needed to reach ideal goals. The clinical benefits of lipid management are often clear. The fact that this man is doing well 23 years following his initial visit is most certainly related to management of his diabetes and lipid disorder.

II. Young man with a history of recurrent pancreatitis and marked elevation of triglyceride-rich lipoproteins managed with alcohol restriction.

History: A 35-year-old man with a history of two episodes of pancreatitis in the previous 18 months was referred for control of marked hypertriglyceridemia. There was no history of cardiovascular disease. There was no family history of early onset coronary artery disease or pancreatitis. He was an imprecise historian but admitted to consuming unknown quantities of alcohol on a regular basis. Physical examination was normal. He was not obese. See the table for response to treatment. Initial visit is highlighted in yellow.

Date	TC	TG	HDL	LDL	non-HDL	non-HDL/ HDL	FBS	Hb-A1c	Wt	Med
12-81	503	3940	30		473	15.8				none
02-82	184	269								none
06-84	200	382	46	78	154	3.3	92			none
07-84	171	282	38	77	133	3.5				none

TC - total cholesterol, TG - total triglycerides, HDL - high-density lipoprotein cholesterol, LDL - low-density lipoprotein cholesterol, non-HDL - cholesterol not in HDL, non-HDL/HDL - ratio non-HDL/HDL cholesterol, FBS - fasting blood sugar, HbA1c - hemoglobin A1c, Wt - weight, Med - lipid medications

Follow-up: Initial studies confirmed those noted prior to referral (marked elevation of triglyceride-rich lipoproteins). He was asked to refrain from drinking alcohol for two weeks and this was successfully done February 1982. One notes a marked reduction in triglyceride-rich lipoproteins with this single maneuver. The patient moved from the Bay Area but returned some two years later in a new relationship now on a lower-fat diet and drinking "only" three beers daily.

One clause of pancreatitis is marked elevation of triglyceride-rich lipoproteins. The pancreas produces enzymes that hydrolyze (digest) triglycerides. Following a meal these enzymes are secreted into the intestine where digestion takes place. In the advent of high blood triglycerides this digestion may occur within the pancreas leading to potentially life threatening pancreatitis.

Prevention is controlling blood triglyceride levels. Blood levels are a function of triglyceride production and removal. Alcohol ingestion leads to increased production of triglyceride-rich lipoproteins from the liver. When clearance systems are overwhelmed, marked hypertriglyceridemia can occur. Alcohol abstinence or restriction often leads to improvement. This is shown dramatically above. This man's pancreatitis was secondary to high triglyceride levels and recurrence was obviated by near normalization. One notes that his triglycerides with reduced alcohol intake are not normal.

Elevated triglyceride-rich lipoproteins not only lead to increased risk of heart attack and stroke but also pancreatitis. Control obviates these problems. Alcohol intake in part governs lipoprotein levels.

III: Primary prevention in a woman with elevated LDL cholesterol managed with statin alone.

History: 57-year-old woman self referred for lipid evaluation. She was well with no major medical problems. She was a non-smoker and walked for exercise without physical limitation. Family history was pertinent in that her mother at age 60 had emergent angioplasty and later coronary artery bypass graft surgery, maternal grandfather died of a myocardial infarction at age 55, and her younger brother recently suffered a myocardial infarction at age 52 (former smoker). Physical examination was normal. She was not obese. Diet was thoughtful with rare alcohol use. See the table for response to treatment. Initial visit is highlighted in yellow.

Date	TC	TG	HDL	LDL	non-HDL	non-HDL/HDL	FBS	Hb-A1c	Wt	Med
06-05	273	111	95	156	178	1.9	80			
05-06	243	57	69	163	174	2.5	87	5.7		
07-06	174	48	80	84	94	1.2				S 40
12-06	161	59	71	78	90	1.3				S 40
07-08	170	79	74	80	96	1.3				S 40

TC - total cholesterol, TG - total triglycerides, HDL - high-density lipoprotein cholesterol, LDL - low-density lipoprotein cholesterol, non-HDL - cholesterol not in HDL, non-HDL/HDL - ratio non-HDL/HDL cholesterol, FBS - fasting blood sugar, HbA1c - hemoglobin A1c, Wt - weight, Med - lipid medications, S 40 - simvastatin 40 mg.

Follow-up: This woman has an increased risk of a vascular event based on elevated LDL-cholesterol and a clear family history of heart attack. Her lipids were normalized with simvastatin 40 mg alone.

This is an example of primary prevention - reducing risk by normalizing lipid levels. She had a better than average response to simvastatin 40 mg. Response to statin prescription is quite variable. Group responses are predictable, but individual response need be assessed. Medication(s) and dosages need be titrated based on response.

IV: Primary prevention in man with mixed dyslipidemia managed with niacin alone.

History. Patient was a 43-year-old married man self referred for elevated cholesterol. He was a nonsmoker, a vegetarian recent 20 years, and avid bicyclist. He had not gained weight since age 20. His father discontinued cigarettes age 44, had his first myocardial infarction age 48, and died of a recurrent MI age 61. His mother was well. Other family history was not available. Physical examination was normal. He was not obese. See the table for response to treatment. Initial visit is highlighted in yellow.

Date	TC	TG	HDL	LDL	non-HDL	non-HDL/ HDL	FBS	Hb-A1c	Wt	Med
11-96	244	393	37	128	207	5.6			189	
01-97	262	258	48	162	214	4.5	83			
03-97	189	123	57	107	132	2.3				N 3.0
03-98	171	119	69	78	102	1.5	81			N 3.0
09-03	145	66	72	60	73	1.0				N 3.0
08-07	141	45	79	53	62	0.8			187	N 3.0

TC - total cholesterol, TG - total triglycerides, HDL - high-density lipoprotein cholesterol, LDL - low-density lipoprotein cholesterol, non-HDL - cholesterol not in HDL, non-HDL/HDL - ratio non-HDL/HDL cholesterol, FBS - fasting blood sugar, HbA1c - hemoglobin A1c, Wt - weight, Med - lipid medications, N 3.0 - niacin 3.0 g.

Followup: He was at increased risk for a cardiovascular event based on family history and mixed dyslipidemia - elevated VLDL and LDL and low HDL. With the institution of rapid-release niacin alone, we see a normalization of VLDL, LDL, and HDL concentrations. With this single maneuver, he has achieved ideal lipid goals with a ratio of non-HDL to HDL cholesterol approaching unity. He continues to be asymptomatic.

This is another example of primary prevention – where the object of treatment is to normalize the lipoprotein abnormality. Regular release niacin can be administered with monitoring at doses up to 6 grams. Niacin uniquely lowers all atherogenic lipoproteins and increases the concentration of HDL. With niacin, HDL concentrations often continue to increase in the early months of treatment.

V: Primary prevention in a man with severe combined hyperlipidemia managed with a combination regimen, niacin and a statin.

History: Patient was a 49-year-old man when referred for management of his elevated lipids. There was no history of symptomatic cardiovascular disease. He had gained 40 pounds since age 20. He was a former pipe and cigar smoker. He drank an occasional glass of wine. There was a strong paternal history of type 2 diabetes and his father died of a myocardial infarction at age 60. On physical examination he was moderately obese and had a mild corneal arcus. He was on no medications at referral. A recent cardiac exercise test was normal. See the table for response to treatment. Initial visit is highlighted in yellow.

Date	TC	TG	HDL	LDL	non-HDL	non-HDL/HDL	FBS	Hb-A1c	Wt	Med
09-87	353	332	44	243	309	7.0	91	7.0	177	
01-88	284	233	41	196	243	5.9	91		166	
04-88	232	167								N 1.5
05-88	217	83	68	132	149	2.2	94		165	N 3.0
05-91	223	93	69	135	154	2.2			182	N 3.0
09-91	163	62	72	79	91	1.3	83		186	N 3.0, L 20
06-99	170	114	80	67	90	1.1			192	N 3.0, S 40
10-06	155	77	87	53	68	0.8	99	6.3	170	N 3.0, S 40
07-08	156	177	73	48	83	1.1	118	6.7	175	N 3.0, S 40

TC - total cholesterol, TG - total triglycerides, HDL - high-density lipoprotein cholesterol, LDL - low-density lipoprotein cholesterol, non-HDL - cholesterol not in HDL, non-HDL/HDL - ratio non-HDL/HDL cholesterol, FBS - fasting blood sugar, HbA1c - hemoglobin A1c, Wt - weight, Med - lipid medications, N 3.0 - niacin 3g, L 20 - lovastatin 20 mg, S 40 -simvastatin 40 mg.

Follow-up: This man was at increased risk for cardiovascular disease based on family history and severe hyperlipidemia (elevated VLDL and LDL). He improved with weight loss, but with residual marked elevations of VLDL and LDL. He had a clear response to the institution of IR niacin first 1.5 g, then 3g daily (decrement in both VLDL and LDL and increase in HDL - note change in ratio of non-HDL to HDL cholesterol). This is the typical response seen with institution of moderate dose niacin. Residual elevations improved with the institution of statin, first lovastatin 20 mg and subsequently simvastatin 40 mg. Lipids have been similar with his variations in weight, over now 21 years. He has remained asymptomatic, now aged 70 years.

This case is a demonstration of the effects of niacin in combination with statin at moderate dose. He has very likely benefited from this regimen, this statement based on his family history and likelihood of such an event in an individual with such marked hyperlipidemia. While blood glucose has risen modestly, this likely relates to age in a man who is obese. Niacin can be used safely in such individuals. The institution of niacin therapy can increase blood glucose in the short-term but most often glucose returns to baseline. If diabetes does develop, treatment is combined hygienic and pharmacologic measures.

VI: Primary prevention in a woman with elevated IDL and VLDL (dysbetalipoproteinemia) managed with a combination regimen, statin, niacin, and omega-3 fatty acids.

History: Patient was a 60-year-old woman when referred for lipid management. Other pertinent problems included a history of surgery, chest radiation and chemotherapy for breast cancer diagnosed age 48 - now cancer free. She was told of elevated lipids during chemotherapy. She discontinued cigarette smoking age 27, consumed alcohol rarely, gained 25 pounds as an adult. There was no family history of early coronary heart disease. She was moderately obese with no stigmata noted related to her lipid disorder. See the table for response to treatment. Initial visit is highlighted in yellow.

Date	TC	TG	HDL	LDL	non-HDL	non-HDL/ HDL	FBS	Hb-A1c	Wt	Med
09-98	274	459	42		232	5.5	89			
02-03	286	334	58	161	228	3.9	85	6.1	158	S 40
04-03	184	466	44		140	3.2			158	S 40, n-3 3
06-03	121	157	58	32	63	1.1			157	S 40, n-3 3, N 1.5
11-03	125	99	64	41	61	1.0	91		158	S 40, n-3 3, N 1.0
12-06	113	88	73	22	40	0.5	97		153	S 40, n-3 3, N 1.0
06-08	131	114	76	32	55	0.7			149	S 10, n-3 3, N 1.0

TC - total cholesterol, TG - total triglycerides, HDL - high-density lipoprotein cholesterol, LDL - low-density lipoprotein cholesterol, non-HDL - cholesterol not in HDL, non-HDL/HDL - ratio non-HDL/HDL cholesterol, FBS - fasting blood sugar, HbA1c - hemoglobin A1c, Wt - weight, Med - lipid medications, S 40 - simvastatin 40 mg, n-3 3 - EPA + DHA 3g, N 1.0 - niacin 1 g.

Follow-up: This woman was first seen in February 2003. Special studies conducted at that time demonstrated that her primary lipid abnormality was elevation of her intermediate-density lipoproteins. This is characteristic of a recessively inherited disorder called dysbetalipoproteinemia. This suspicion was confirmed with genotyping and demonstration that she was apo E2 – E2 (inherited E2 from each parent – common E genotype is E3). These specialized tests are available at selected commercial laboratories. Based on historic records compared to initial visit (Table), she had no response to simvastatin 40 mg given as a single agent. The omega-3 fatty acids - EPA plus DHA are often useful in treating this disorder. She had a partial response to adding the omega-3 fatty acids. We have only one data set on combination simvastatin and n-3 fatty acids. Niacin was added with a remarkable response observed at low dosage. With purposeful weight loss her lipids improved allowing a decrease in simvastatin dosage to 10 mg. She continues now with ideal lipids. Clinically she was evaluated for complaints of mild breathlessness. At coronary angiogram she was found to have diffusely narrowed coronary vessels and two stents were placed. The cardiologist who did the procedure felt this diffuse narrowing may be attributed to her prior chest irradiation. She felt no improvement in mild breathlessness with exercise with stent placement.

Dysbetalipoproteinemia often responds partially to hygienic measures and one of the three agents used above. To achieve ideal lipids in this case, three drugs, albeit at low dosage, were necessary to achieve optimal values. One should note the real increase in HDL with the institution of niacin at very-low dosage (not typical). Dysbetalipoproteinemia, when lipids are elevated, is atherogenic and it is likely that this woman will benefit in the long-term from ideal lipid control. She likely has another genetic disorder leading to increased production of VLDL and thus IDL. Most people who inherit apo E2-E2 are not hyperlipidemic.

VII: Primary prevention in a man with combined hyperlipidemia managed with a combination regimen, niacin and statin.

History: A 63-year-old man was referred for lipid management. He was well with no other important medical problems. He discontinued cigarettes at age 26, his food choice was thoughtful and alcohol intake was limited to five glasses of wine weekly. He exercised regularly with no physical limitation. His father, a non-smoker, sustained a myocardial infarction at age 43, later a stroke, was diagnosed with type 2 diabetes age 60, and died of a stroke age 65. Patient's hyperlipidemia was noted in the recent decade and was reportedly unresponsive to lovastatin and his current regimen of ezetimibe. See the table for response to treatment. Initial visit is highlighted in yellow.

Date	TC	TG	HDL	LDL	non-HDL	non-HDL/HDL	FBS	Hb-A1c	Wt	Med
01-04	243	280	46	141	197	4.3	94			Ez 10
03-04	200	229	47	107	153	3.3			168	Ez 10
07-04	235	182	60	139	175	2.9				N 3.0
11-04	194	76	73	106	121	1.7	92			N 4.0
04-05	144	71	64	66	80	1.3				N 4.0, L 20
11-08	127	59	69	46	58	0.8	90	5.6	169	N 4.0, L 20

TC - total cholesterol, TG - total triglycerides, HDL - high-density lipoprotein cholesterol, LDL - low-density lipoprotein cholesterol, non-HDL - cholesterol not in HDL, non-HDL/HDL - ratio non-HDL/HDL cholesterol, FBS - fasting blood sugar, HbA1c - hemoglobin A1c, Wt - weight, Med - lipid medications, Ez 10 - ezetimibe 10 mg, N 4.0 - niacin 4.0 g, L 20 - lovastatin 20 mg.

Follow-up: This man was at increased risk for a vascular event based in on modest combined abnormalities of VLDL and LDL and a paternal history of myocardial infarction and stroke. His dyslipidemia was unresponsive to trials of lovastatin and ezetimibe. Lipid concentrations were quite variable and responded modestly to niacin 4 g (increased HDL and decreased non-HDL cholesterol). With the addition of lovastatin 20 mg, he moved to ideal goals with ratio of non-HDL to HDL cholesterol near unity.

In a primary prevention setting, risk reduction is achieved by normalizing lipoproteins. Of interest in this example, historically he had not responded to lovastatin, and his response to niacin was unimpressive at 3g and modest only at 4 g. However, goal was achieved with these two agents in combination. On occasion, one sees a response to a combination, where response to the individual components is not impressive.

VIII: Primary prevention in a man with markedly elevated triglyceride-rich lipoproteins managed initially with omega-3 fatty acids.

History: 35-year-old man referred for lipid management. He was basically well and active in sports. He was a non-smoker. He consumed 14-20 alcohol containing beverages weekly. Elevated lipids were noted on earlier routine testing. There was no known family history of elevated lipids or premature cardiovascular disease. Niaspan 1g was initiated by the referring physician. See the table for response to treatment. Initial visit is highlighted in yellow.

Date	TC	TG	HDL	LDL	non-HDL	non-HDL/HDL	FBS	Hb-A1c	Wt	Med
07-02	298	1570	20		278	13.9				
10-02	307	810	35		272	7.8	91	5.4	237	N 1.0
11-02	192	214	37	112	155	4.2			236	N 1.0
08-03	277	1390	17		260	15.3				
11-03	183	235	30	106	153	5.1			237	n-3 3.0
10-04	234	341	35	131	199	5.7	105		228	n-3 3.0
11-04	137	121	40	73	97	2.4				n-3 3.0, Vy 10/20
01-05	116	129	40	50	76	1.9			231	n-3 3.0, Vy 10/20

TC - total cholesterol, TG - total triglycerides, HDL - high-density lipoprotein cholesterol, LDL - low-density lipoprotein cholesterol, non-HDL - cholesterol not in HDL, non-HDL/HDL - ratio non-HDL/HDL cholesterol, FBS - fasting blood sugar, HbA1c - hemoglobin A1c, Wt - weight, Med - lipid medications, N 1.0 - niacin 1g, n-3 3.0 - EPA +DHA 3g, Vy 10/20 - Vytorin 10/20 (ezetimibe 10 mg/simvastatin 20 mg)

Follow-up: This man was at increased risk for a vascular event based on marked elevation of his triglyceride-rich lipoproteins. Note that his LDL cholesterol can not be estimated with a total triglyceride greater than 400. When first seen in October 2002, it was suggested that he discontinue all alcoholic beverages. One month later his lipids were markedly improved but not normal. By August 2003, he had returned to his old alcohol habits and lipid profile. He initiated omega-3 fatty acids at 3g of EPA plus DHA with clear improvement in his lipids. Again he was not at goal. Near ideal goals were achieved with the addition of Vytorin 10/20.

It is very likely that this man has a decreased ability to process (clear) triglyceride-rich lipoproteins (recessive inheritance is the usual). VLDL concentrations are secondarily strongly determined by other factors such as habitual alcohol intake, excess calories, stress or diabetes control. The omega-3 fatty acids EPA and DHA when consumed in gram quantities often reduce the production of VLDL. Normalizing lipoproteins in people with a clearance defect may require the addition of niacin, or a statin or all agents in combination.

IX: Primary prevention in a man with severe elevations of LDL and VLDL - managed with a combination regimen, statin, niacin and ezetimibe.

History: 54-year-old man self referred for treatment of combined hyperlipidemia. He was a former cigarette smoker, having stopped at age 20. He drank alcohol modestly. Food choices were near ideal. He was not obese. Family history was pertinent in that his father (a cigarette smoker) had his first myocardial infarction at age 45 and succumbed to a MI at age 58. Two paternal aunts died of myocardial infarctions in their 60s. The patient was well. See the table for response to treatment. Initial visit is highlighted in yellow.

Date	TC	TG	HDL	LDL	non-HDL	non-HDL/HDL	FBS	Hb-A1c	Wt	Med
1-85	429	320	63	302	366	5.8				
2-92	194	151	75	89	119	1.6	75		159	L 60,C 30g
11-98	191	87	84	90	107	1.3	95		160	S 40, N 3.0
10-02	188	43	99	80	89	0.9	81		162	S 40, N 3.0
4-03	129	42	83	38	46	0.6		5.2		S40, N 3.0, Ez 10
2-04	151	60	91	48	60	0.7			162	S 40, N 3.0, Ez 10
6-08	154	50	79	65	75	0.9	92	5.6	159	Vy 10/40, N 3.0
1-09	147	44	75	63	72	1.0	89	5.6	166	Vy 10/40, N 3.0

TC - total cholesterol, TG - total triglycerides, HDL - high-density lipoprotein cholesterol, LDL - low-density lipoprotein cholesterol, non-HDL - cholesterol not in HDL, non-HDL/HDL - ratio non-HDL/HDL cholesterol, FBS - fasting blood sugar, HbA1c - hemoglobin A1c, Wt - weight, Med - lipid medications, L 60 - lovastatin 60 mg, C 30 - colestipol 30 g, S 40 - simvastatin 40 mg, N 3.0 - niacin 3g, Ez 10 - ezetimibe 10 mg, Vy 10/40 - Vytorin 10/40 (ezetimibe 10 mg/simvastatin 40 mg)

Follow-up: This man was clearly at increased risk of an early vascular event based on marked hyperlipidemia and family history of early death. At presentation, he had marked elevation of LDL and elevated VLDL. He was first treated in the context of a drug protocol receiving what was later to be marketed as lovastatin (Mevacor). Lovastatin, the first statin, was FDA approved and marketed in 1987. Later he achieved near normalization of his lipids with a combination of lovastatin and the bile acid sequestrant, colestipol. The regimen changed over the years sequentially to a combination of simvastatin and niacin, later to simvastatin, niacin and ezetimibe, and currently Vytorin 10/40 (combined simvastatin and ezetimibe) and niacin. His ratio of non-HDL to HDL is ideal (unity).

This is another example of primary prevention prompted by self assessment of his dire family history of early coronary death. He continues to do well with no cardiac problems now 24 years later and aged 78 years.

X: Primary and now secondary prevention in a man with modestly elevated LDL and low HDL cholesterol managed with a combination regimen, statin and niacin.

History: 74-year-old man was self referred February 2004 for management of hyperlipidemia. He was well and active physically. He was a life-long nonsmoker. He drank two servings of alcohol weekly and was on a thoughtful diet. He was diagnosed with hypertension in 1998 which had been controlled with medications. His father died at age 36 from an infection. His mother died at age 91 of congestive heart failure. There was no definite family history of early coronary artery disease but his paternal grandfather and uncle died in their 60s of unknown cause. Medications included two agents for hypertension control and he had been placed on atorvastatin a month prior to self-referral. See the table for response to treatment. Initial visit is highlighted in yellow.

Date	TC	TG	HDL	LDL	non-HDL	non-HDL/HDL	FBS	Hb-A1c	Wt	Med
08-03	186	108	42	122	144	3.4				
02-04	133	45	36	88	97	2.7	93		168	A 10
04-04	130	49	41	79	89	2.2			167	A 20
06-04	133	22	70	59	63	0.9			154	A 20, N 1.5
12-04	124	32	80	38	44	0.6			159	A 20, N 1.5
01-08	126	42	82	36	44	0.5	106	5.8	159	A 20, N 1.5
11-08	128	40	83	37	45	0.5	98	5.9	159	A 10, N 1.5

TC - total cholesterol, TG - total triglycerides, HDL - high-density lipoprotein cholesterol, LDL - low-density lipoprotein cholesterol, non-HDL - cholesterol not in HDL, non-HDL/HDL - ratio non-HDL/HDL cholesterol, FBS - fasting blood sugar, HbA1c - hemoglobin A1c, Wt - weight, Med - lipid medications, A 10 or 20 - atorvastatin 10 or 20 mg, N 1.5 - niacin 1.5g

Follow-up: This man presented with modest increase of non-HDL cholesterol and low HDL cholesterol. Additional studies were recommended including LP(a) - normal, hs CRP <0.5 (low risk) and a computed tomography (X-ray) of the heart to look for coronary artery calcification. In most individuals, long established plaque in the heart arteries prompts visible calcification. This study demonstrated 28 lesions in all three coronary arteries. With this information, atorvastatin dosage was first increased and then niacin was added to the regimen. One notes the addition of niacin did not change the total cholesterol but in fact lead to a remarkable redistribution of cholesterol, increasing HDL and decreasing the non-HDL lipoproteins dramatically (ratio decrease from >2 to<1). This occurred at a very low dosage (1.5 g daily). In the summer of 2005, he developed "panic attacks." He was referred to a cardiologist and following PTCA his panic attacks resolved. He now feels well. Of interest is his move to a four-story walk-up and enjoying the stairs.

This is an example of first primary and now secondary prevention. The decision to treat with lipid lowering agents is not always clear. Use of coronary artery calcification screening is a good noninvasive test along with exercise ECHO to determine the extent of current trouble. If there is evidence of coronary artery or other large vessel disease, I would suggest aggressive treatment. It is likely this man will do well.

XI: Secondary prevention in a woman with elevated LDL and Lp(a) who presented with right lower extremity claudication (exercise induced leg cramping) managed with a combination regimen, statin and niacin.

History: 70-year-old woman was referred by her vascular surgeon for lipid management. She had a history of hypertension diagnosed at age 60 but was otherwise well when she developed leg cramps with walking in the spring of 2002. She saw a vascular surgeon and underwent angioplasty but with only temporary relief. Her second vascular surgeon suggested referral. She discontinued cigarettes at age 50 and consumed alcohol only rarely. Food choice was good. Family history revealed that her father died at age 53 of a stroke but her mother was well at age 95. Of concern was the fact that her sister died suddenly at age 45 of unclear cause. Prior to referral she was given atorvastatin in May of 2002 and in June, Niaspan 500 mg. This was discontinued because of flushing and little recorded response. See the table for response to treatment. Initial visit is highlighted in yellow.

Date	TC	TG	HDL	LDL	non-HDL	non-HDL/HDL	FBS	Hb-A1c	Wt	Med
02-98	228	59	54	162	174	3.2	105			
10-01	202	23	55	142	147	2.7				
11-02	177	49	69	98	108	1.6	88		124	A 10
01-03	201	41	127	66	74	0.6			127	A 10, N 1.5
02-04	167	41	95	64	72	0.8	120	5.7	133	A 10, N 3.0
03-07	159	53	100	48	59	0.6	122	5.7	123	A 10, N 3.0
11-08	160	61	94	54	66	0.7	100	6.1	123	A 10, N 3.0

TC - total cholesterol, TG - total triglycerides, HDL - high-density lipoprotein cholesterol, LDL - low-density lipoprotein cholesterol, non-HDL - cholesterol not in HDL, non-HDL/HDL - ratio non-HDL/HDL cholesterol, FBS - fasting blood sugar, HbA1c - hemoglobin A1c, Wt - weight, Med - lipid medications, A 10 - atorvastatin 10 mg, N 3.0 - niacin 3g

Follow-up: This woman had modest but definite elevation of LDL and non-HDL cholesterol and variably increased fasting blood glucose. Additional studies revealed a clearly elevated Lp(a) and hs CRP of 1.0 (borderline between low or moderate risk). Regular niacin was added with titration to 1.5 g with clear improvement in her lipids. With advancement to 3 g, Lp(a) normalized. She underwent successful angioplasty and stenting of her right internal iliac artery December 2002 and subsequently has been asymptomatic. She leads a very active life and with her husband (case above) resides in a four-story walk-up in San Francisco, trekking the stairs multiple times daily.

In this example it was important to look beyond the standard lipid measures. Recent large population studies have unequivocally established elevated Lp(a) as an important, independent risk factor for atherosclerotic disease. The addition of niacin to the regimen resulted in a radical redistribution of cholesterol between non-HDL and HDL cholesterol (dramatic fall in the ratio) and normalization of Lp(a). Niacin is the only agent which lowers Lp(a) and in this case it was normalized. Measurements of glucose metabolism (fasting glucose and hemoglobin A1c [reflecting average glucose levels]) may increase with niacin prescription, prompting further effort in establishing weight reduction and restriction of simple carbohydrates. If blood glucose can not be controlled with hygienic efforts, pharmacologic measures may be required. In clinical trials of diabetic patients with dyslipidemia, niacin prescription has resulted in reduction of cardiovascular events. This woman is very likely to continue doing well.

XII: Secondary prevention in a man with elevated LDL and VLDL managed finally with a combination regimen, statin, niacin, and ezetimibe.

History: 51-year-old man was referred for management of his marked hyperlipidemia. He survived coronary artery bypass graft surgery at age 47. When he presented, he felt well leading an active life. He was a lifelong non-smoker and drank alcohol moderately. He had a family history of aggressive coronary artery disease with both his brother and father dying of myocardial infarctions at an early age. He had no other significant medical problems. He was obese. See the table for response to treatment. Initial visit is highlighted in yellow.

Date	TC	TG	HDL	LDL	non-HDL	non-HDL/HDL	FBS	Hb-A1c	Wt	Med
01-85	438	205	53	344	385	7.3				
10-98	201	104	68	112	133	2.0	111		231	A 40, N 4.5
04-00	191	111	79	90	112	1.4	90		228	A 40, N 4.5
04-01	163	108	71	70	92	1.3	114	5.6	223	A 40, N 4.5, W 3.8
04-03	174	119	69	81	105	1.5			213	A 40, N 4.5, W 3.8
06-03	133	92	72	43	61	0.8	93		214	A 40, N4.5, Ez 10
05-04	151	78	63	72	88	1.4			220	A 40, N4.5, Ez 10
11-08	129	70	65	50	64	1.0	93	6.3	216	A 40, N4.5, Ez 10

TC - total cholesterol, TG - total triglycerides, HDL - high-density lipoprotein cholesterol, LDL - low-density lipoprotein cholesterol, non-HDL - cholesterol not in HDL, non-HDL/HDL - ratio non-HDL/HDL cholesterol, FBS - fasting blood sugar, HbA1c - hemoglobin A1c, Wt - weight, Med - lipid medications; A 40 - atorvastatin 40 mg, N 4.5 - niacin 4.5g, W 3.8 - WelChol (colesevelam) 3.8g, Ez 10 - ezetimibe 10 mg.

Follow-up: This man presented 4 years after coronary artery bypass surgery for management of severe hyperlipoproteinemia. His treatment and clinical course is a testimony to advances in lipoprotein management. He first entered a research protocol which included use of what was later marketed as lovastatin (Fall 1987). Initial treatment included statin and resin. Later we pick up the story in October of 1998 with a regimen of statin and niacin. The addition of colesevelam, a bile acids sequestrant, showed marginal benefit. The addition of ezetimibe lead to further significant reductions in LDL and non-HDL cholesterol. This man has had a dramatic percent reduction in non-HDL cholesterol and now is at ideal levels. He continues to do well clinically now 27 years post coronary artery bypass surgery and aged 75 years.

This is an example of secondary prevention in a man with severe coronary artery disease and marked hyperlipidemia. Examples such as this one demonstrate the benefit of normalizing lipids.

FINAL THOUGHTS

This book was written to provide the reader basic information needed to understand the risks conferred by abnormalities in blood lipoproteins and the tools currently available to reduce lipoprotein related risk.

Moving to goal requires the concerted efforts of the interested patient in concert with his or her physician. Most physicians understand the content discussed in this Little Book but not all are equally skilled in treating lipid disorders. Least understood is the value of immediate release niacin, experience with its prescription, and the dose range available to the clinician. The assistance of a specialist may be sought when necessary.

 The examples provided illustrate response with proper choice of available tools. Important to note, however, is that any given manipulation leads to a variety of responses mandating that the regimen used need to be modified to achieve goal. Record keeping by both the patient and physician is critical. Results are easily quantitated based on simple blood test determinations.

REFERENCES AND COVER CREDITS

1. Havel RJ. in "Pathophysiology: The Biological Principles of Disease" (edited by) Lloyd H. Smith, Jr., Samuel O. Thier. Published in 1981 as Volume I of "International Textbook of Medicine" (W.B. Saunders Co.). Adapted from Table 20 (page 600), chapter entitled "Metabolism and Nutrition" (Felig, Havel and Smith).

2. Lloyd-Jones DM, Wilson PWF, Larson MG, Leip E, Beiser A, D'Agostino RB, Cleeman JI, Levy D. . Lifetime risk of coronary heart disease by cholesterol levels at selected ages. Archives of Internal Medicine (2003) vol. 163 pp. 1966-72

3. Stamler J, Daviglus ML, Garside DB, Dyer AR, Greenland P, Neaton JD. Relationship of baseline serum cholesterol levels in 3 large cohorts of younger men to long-term coronary, cardiovascular, and all-cause mortality and to longevity. Journal of the American Medical Association (2000) vol. 284 pp. 311-8

4. Lloyd-Jones DM, Leip EP, Larson MG, D'Agostino RB, Beiser A, Wilson PWF, Wolf PA, and Levy D. Prediction of lifetime risk for cardiovascular disease by risk factor burden at 50 years of age. Circulation (2006) vol. 113 pp. 791-8

5. Lloyd-Jones DM, Dyer AR, Wang R, Daviglus ML, Greenland P. Risk factor burden in middle age and lifetime risks for cardiovascular and non-cardiovascular death (Chicago Heart Association Detection Project in Industry). American Journal of Cardiology (2007) vol. 99 pp. 535-40

6. National Cholesterol Education Program (NCEP) Expert Panel on Detection, Evaluation, and Treatment of High Blood Cholesterol in Adults (Adult Treatment Panel III). Third Report of the National Cholesterol Education Program (NCEP) Expert Panel on Detection, Evaluation, and Treatment of High Blood Cholesterol in Adults (Adult Treatment Panel III) final report. Circulation (2002) vol. 106 pp. 3143-421

7. Grundy SM; Cleeman JI; Bairey Merz CN; Brewer, Jr HB; Clark LT; Hunninghake DB; Pasternak RC; Smith, Jr SC; Stone NJ; for the Coordinating Committee of the National Cholesterol Education Program. Implications of recent clinical trials for the National Cholesterol Education Program Adult Treatment Panel III guidelines. Circulation (2004) vol. 110 pp. 227-39

8. Cohen JC; Boerwinkle E; Mosley, Jr TH; Hobbs HH. Sequence variations in PCSK9, low LDL, and protection against coronary heart disease. New England Journal of Medicine (2006) vol. 354 pp. 1264-72

COVER CREDIT AND REFERENCE

1. Front cover: Electron micrograph kindly provided by R. L. Hamilton, Ph.D., University of California, San Francisco. Lipoproteins were separated from blood plasma utilizing the preparative ultracentrifuge. Magnification is listed beneath the figure. Pictures were taken after negative staining and lipoproteins are visualized on a dark background.

2. Back cover reference: Friedewald WT, Levy RI, Fredrickson DS. Estimation of the concentration of low-density lipoprotein cholesterol in plasma, without the use of the preparative ultracentrifuge. Clinical Chemistry (1972) vol. 18 pp. 499-502

APPENDIX I - METHODS FOR INITIATING NIACIN

Standard Method:

1. Use regular niacin, not a timed-release or sustained-release preparation. I strongly recommend an encapsulated product specifically the TWINLAB brand. TWINLAB niacin is available as 500-mg and 1000-mg capsules and can be purchased at most health food stores and on the Internet. The label reads "NIACIN (B-3) CAPS." Other TWINLAB niacin products are not recommended. 100-mg tablets are used to initiate therapy. The brand of niacin 100-mg tablet is not critical. Niacin 500-mg tablets are readily available over the counter. While not recommended, they may be suitable. Other encapsulated products may be available.

2. Everyone beginning niacin will have a characteristic "flush" reaction with initial doses. Occasionally the response will include itching and a transient rash. This reaction will be tempered by taking the dose just after a meal, and avoiding hot drinks and alcohol initially. Taking aspirin just before the dose decreases the intensity of the flush.

3. Take the first dose at home and at a quiet time to avoid any possible confusion commensurate with a strong flush and a social interaction.

4. To begin, purchase a bottle of 100-mg niacin tablets (100 tablets), and a bottle of 500-mg TWINLAB capsules (TWINLAB NIACIN (B-3) CAPS [100 capsules/bottle]).

5. Begin with 100-mg tablets, one three times daily with meals for a week, then advance the dosage to 2 tablets (200 mg) three times daily for seven days, followed by three (300 mg) three times a day. When this bottle is exhausted, move to the 500 mg capsule preparation, one capsule 3 times daily with meals (500-mg thrice daily).

This regimen is summarized below:

Day 1-7	100 mg (1 tablet)	3 times daily with meals
Day 8-14	200 mg (2 tablets)	3 times daily with meals
Day 15-18	300 mg (3 tablets)	3 times daily with meals
Day 19-	500 mg (1 capsule)	3 times daily with meals

6. Maintain the dosage of 500 mg three times daily for several weeks until repeat blood testing, after which you will likely be asked to increase the dose.

7. The symptoms associated with niacin initiation always decrease with time and as the dosage increases often subside entirely.

8. Niacin is a well studied agent. When used properly, its use has been demonstrated to reduce the risk of heart attack and increase lifespan.

Accelerated Method:

The accelerated method is suitable for the niacin veteran who for any reason, had interrupted his established niacin regimen. It however is adaptable for niacin initiation.

1. Prior to a meal, take one gram (1000 mg) of aspirin (three standard dosage aspirin tablets) with a large glass of water.

2. After the meal, return to the previously established niacin regimen. For example, if the established regimen was 1000 mg of niacin taken three times daily - take 1000 mg as a first dose.

3. Following numbers 1 and 2 above, simply proceed with the previously established niacin regimen. In the example above, proceed with niacin 1000 g three times daily.

4. If the accelerated method is used to initiate niacin therapy, take 1000 mg of aspirin followed by a niacin 500 mg initially and then 3 times daily. Subsequent adjustment of niacin dosage follows laboratory assessment of response (lipids and safety data).

5. In the author's limited experience with the accelerated method, there is little to no associated flush with simply returning to a standard regimen following initial coverage with 1000 mg of aspirin.

APPENDIX II - THE PASSING LANE

Passing lane - Basic Physiology: Cholesterol and triglycerides are both water insoluble and unlike sugars, proteins and salts which are water soluble, they can not be carried dissolved in the blood compartment. The solution is the construction of lipoproteins - spherical particles which transport cholesterol and triglycerides within their core. Six families of lipoproteins are defined. All but one, the smallest and heaviest (HDL), when elevated can contribute to blood vessel disease. Basic measurements include the blood total cholesterol and cholesterol in HDL. The cholesterol not in HDL (non-HDL cholesterol - difference between total and HDL cholesterol) and ratios derived from these measurements (non-HDL cholesterol to HDL cholesterol) can be calculated. Additional measurements (fasted triglycerides, lipoprotein(a), and others) better assess risk of potential blood vessel and other lipid driven problems.

Passing lane - Risk Assessment: Risk assessment is an imprecise art. Standard recommendations emphasize short-term (10-year) risk, but information describing lifetime risk is also available. Following review of large data sets, it is clear that a move to optimal levels of recognized risk factors will improve longevity not only secondary to vascular diseases but from all problems. Individuals with optimal levels of non-HDL cholesterol, HDL cholesterol, blood pressure, smoking status, absence of diabetes, and exercise clearly do better than individuals who have one or more risk factors not optimal or elevated, and much much better then those at high risk. Is it possible to achieve optimal or near optimal levels? Yes, with attention to detail.

Passing lane - How Do Patients Present? Patients present with one or more of many possible lipoprotein abnormalities. While there are general guidelines for treatment, specific treatments require assessment of the lipoprotein disorder.

Passing lane - What Is Goal? Two sets of goals are commonly described. Most people emphasize control of LDL cholesterol but in fact the more inclusive non-HDL cholesterol should be assessed. In managing risk, optimal levels are a non-HDL cholesterol of less than 100. Current data suggests that this goal will reduce short-term and lifetime risk substantially. For individuals not at highest risk, some suggest that goal be a non-HDL cholesterol of less than 130.

Passing lane - How Do We Achieve Goal? Goal requires attention to hygienic measures and frequently a proper medication regimen. Most regimens will include two or more lipid modifying agents. The most under utilized medication is the immediate release preparation of niacin. First-line therapies are those well studied in clinical trials.

Passing lane - What Can We Learn From Patient Stories? Patient stories are examples of what can be achieved with current tools. The reader may benefit from study of a story representing a problem similar to his/her own.

The author, Philip H. Frost, M.D., is a clinical professor of medicine at the University of California, San Francisco (UCSF) in the Department of Medicine and Cardiovascular Research Institute. He is a clinician, clinical investigator, who for more than 40 years has worked in the field of cardiovascular risk reduction. He currently sees patients in Private Practice and at UCSF in the Faculty Lipid Practice. More information can be found at his website www.stopheartattack.org.

www.ingramcontent.com/pod-product-compliance
Lightning Source LLC
Chambersburg PA
CBHW041507280526
45792CB00004B/1162